BISHOP

Water Cycle

Monica Hughes

Heinemann Library
Chicago, Illinois

© 2004 Heinemann Library,
a division of Reed Elsevier Inc.
Chicago, Illinois

Customer Service 888-454-2279
Visit our website at www.heinemannlibrary.com

Editorial: Jilly Attwood, Kate Bellamy
Design: Jo Hinton-Malivoire
Picture research: Ginny Stroud-Lewis, Ruth Blair
Production: Séverine Ribierre

Originated by Dot Gradations Ltd
Printed and bound in China by South China Printing Company

08 07 06 05 04
10 9 8 7 6 5 4 3 2 1

Library of Congress Cataloging-in-Publication Data
The Cataloging-in-Publication Data for this title is on file with the Library of Congress.
ISBN: 1-4034-5880-4 (HC), 1-4034-5886-3 (Pbk.)

Acknowledgments
The author and publishers are grateful to the following for permission to reproduce copyright material:
p. 4 Roberto Contini/Alamy; p. 5 Michael S. Yamashita/Corbis; p. 6 Roy Morsch/Corbis; pp. 7, 14, 20 Getty Images/Photodisc; pp. 8, 29, 30 Tudor Photography; pp. 9, 21 Getty Images; pp. 10–11 Alamy; p. 12 Julie Habel/Corbis; p. 13 Rebecca Emery/Corbis; pp. 15, 19, 25, 26, 27 Corbis; p. 16 Craig Aurness/Corbis; p. 17 John Noble/Wilderness Photo Library; p. 18 Alan Schein/Corbis; p. 24 David Muench/Corbis; p. 28 John Cleare

Cover photograph is reproduced with permission of Corbis.

Our thanks to David Lewin for his assistance in the preparation of this book.

Every effort has been made to contact copyright holders of any material reproduced in this book. Any omissions will be rectified in subsequent printings if notice is given to the Publishers.

The paper used to print this book comes from sustainable resources.

Contents

Some words are shown in bold, **like this.**
You can find out what they mean by looking
in the glossary.

Nature's Patterns

Nature is always changing. Many of the changes in nature follow a **pattern.** This means that they happen over and over again.

Every time we turn on a tap we have fresh, clean water.

Water falls from the sky as rain.

Some patterns are called cycles. Cycles go round and round. They have no clear beginning or end. The water cycle is a pattern like this.

5

Water Is Special

Water is a **liquid.** It is all around us in oceans, rivers, lakes, and rain. We can see it and feel it. It moves and flows.

Water is a liquid we can taste and feel.

Water always flows downward.

Water can also change and take different forms. We see these changes and different forms in the water cycle.

7

Water in the Air

Water is all around us even when we cannot see it. Water is in the air as a **gas.** This gas is called **water vapor.**

Steam from a boiling pot disappears into the air.

8

There is less water vapor in cold air. There is more water vapor in warm air. If there is a lot of water vapor in the air, it feels **humid.**

Some people use fans to help cool down when it is humid.

Oceans and Seas

The world's oceans and seas are an important part of the water cycle. Heat from the Sun warms the **surface** of seas and oceans.

When the water becomes warm, it begins to **evaporate.** It changes from a **liquid** into the **gas** called **water vapor.**

Water is evaporating from the surface of the ocean all the time.

Evaporation on Land

Water **evaporates** from streams, rivers, lakes, and ponds. It evaporates from wet roads and sidewalks. It also evaporates from the **surfaces** of plants and trees.

Wet laundry dries as the water in the clothes evaporates.

As water evaporates, it changes into **water vapor**. All over the world, water is evaporating and changing into water vapor.

Water evaporates from puddles, too.

Clouds

When warm air meets cool air, the **water vapor** in the air changes. Then, the water vapor in the air changes from a **gas** to a **liquid.** This is called **condensation.**

Clouds filled with drops of water form over the sea.

The shape and color of a cloud depends on how much water is in it.

Water vapor cools and changes into tiny drops of water. This makes clouds. As the air gets even cooler, the tiny drops join together and fall to earth as rain.

Mountain clouds

The wind blows some clouds toward mountains. Mountains force the clouds higher, where the air is cooler. The clouds get heavy with rain.

Mountain areas often have more rain.

Usually, rain falls on the side of the mountain that faces the sea. The other side has much less rain. The side with less rain is called the rain shadow.

This side of the mountain is the rain shadow. It has less rain and clouds.

Rain

Most rain falls back into the seas and oceans. The rain then begins to **evaporate.** In this way, the water cycle goes round and round.

Rain that falls in towns and cities flows back to the sea.

Heavy rain brings extra water to rivers and waterfalls.

Rain that falls onto the ground soaks into the soil and drains into rivers. Rivers carry the water back to the seas and oceans.

19

Snow

When it is very cold, the water droplets in clouds **freeze.** The droplets form **solid ice crystals.** These join together as snowflakes and fall from the clouds as snow.

Snowflakes are made of ice.

Snow and ice are water in solid form.

When snow **melts,** it changes back to water. Water from the melted snow flows back to the sea.

The complete cycle

The water cycle goes round and round just like a wheel. Water **evaporates** and changes to **water vapor.**

Water vapor rises and cools to form clouds.

The Sun's heat evaporates water from the sea and living things.

22

In cool air, the water vapor in clouds **condenses.** Water falls from the clouds as rain or snow. The water flows down through the ground. It goes into rivers and streams and flows back to the sea.

Clouds are blown by the wind. They contain tiny drops of water that fall as rain or snow.

Water soaks into the ground or flows into rivers and lakes on its way to the sea.

Short Water Cycle

In some places, water **evaporates** and **condenses** in a short time. In rain forests it is hot during the day, so water evaporates very quickly.

In a rain forest, it rains every day. But the water evaporates quickly.

The rain helps plants in the rain forest grow well.

At night it is cool and clouds form. **Water vapor** in the clouds condenses and falls as rain. But the rain quickly evaporates in the heat, and the water cycle keeps going.

Long Water cycle

In very cold places it takes longer for water to **evaporate** and **condense.** Water droplets in clouds **freeze** and fall as snow. Then, the snow changes to ice.

Rivers slowly carry ice to the sea where it melts.

Ice slowly melts when the Sun is shining on it.

In cold places ice does not **melt.** Sometimes, huge pieces of ice move into warmer water. Here they melt and the water evaporates. Then, the water cycle starts all over.

Salty to fresh

The water in the sea is salty. The water in mountain streams is not salty. It is fresh water. When seawater **evaporates,** the salt is left behind.

You can sometimes see salt on your skin after you swim in the sea.

The water in a mountain spring is clean and pure.

When water goes back to the sea, fresh water mixes with salty water. The water changes from fresh to salty.

Salt Test

This simple test shows how water **evaporates.** Place a saucer of salty water by a sunny window. Leave it there for two or three days. What do you think will happen?

The water evaporates, but the salt is left behind in the saucer. This is what happens to the water in seas and oceans.

Glossary

condense when a gas changes into liquid

evaporate when a liquid changes into a gas

freeze when a liquid changes into a solid

gas like air, not a liquid or solid

humid warm and damp

ice crystal small piece of ice

liquid something that flows or can be poured

melt when a solid changes into liquid

pattern something that happens over and over again

solid firm, not liquid

surface area on the top of something

water vapor water as an invisible gas

More Books to Read

Frost, Helen. *The Water Cycle*. Mankato, Minn.: Capstone, 1999.

Hammersmith, Craig. *Rising Up, Falling Down*. Minneapolis, Minn.: Compass Point, 2002.

Robinson, Fay. *Where Do Puddles Go?* Danbury, Conn.: Scholastic Library, 1995.

Ross, Michael Elsohn. *Re-Cycles*. Brookfield, Conn.: Millbrook, 2002.

Index